Math Concept Reader

Putting the World on a Page

by Sarah Mastrianni

Photographs by Russell Pickering

Printed in China

ISBN 13: 978-0-15-360245-0
ISBN 10: 0-15-360245-7

11 12 13 0940 16 15 14
4500492255

Chapter 1:

Stamping Through Australia and Japan

Julia's grandfather is a philatelist. A philatelist is a person who collects stamps. During his travels, Grandfather finds stamps. He saves them as souvenirs. When he receives letters from family and friends, he often keeps the stamps from the envelopes, too. Over the years, he has collected many stamps that way.

Today, Julia's grandfather wants Julia to help him put all of his stamps into one big album. Julia is excited about the project. She enjoys spending time with Grandfather. He tells fascinating stories about how each stamp became part of his collection. Julia and Grandfather plan how they will organize the new album.

Grandfather has a lot of stamp albums. He arranged his smaller stamp albums by country. Over the years, the albums have become disorganized. Now, many stamps are out of place.

Julia thinks arranging the stamps by country is a great idea. First, Julia and Grandfather will need to sort and reorganize the stamps. Then they will place the stamps in the new album. Julia flips through Grandfather's stamp albums. She finds some interesting stamps from Japan. That's a good place to start. Julia begins to arrange the first page of the new stamp album. She makes nine rows of four stamps each.

These stamps are from Japan.

These Australian stamps include birds, flowers, and people.

While Julia arranges stamps from Japan, Grandfather searches through his stamp albums. He locates a number of stamps from Australia. In fact, he finds enough stamps to make six rows of six stamps. Julia looks at the variety of pictures on the stamps. She notices colorful pictures of birds, flowers, and people.

Grandfather points to one stamp in particular. It features a picture of whale. Grandfather knows that Julia loves whales! He tells Julia that at one time Australia issued a series of stamps called "Creatures of Slime." Julia laughs. She tries to imagine what would be pictured on those stamps.

Julia and Grandfather take a moment to look at their page layouts. They need to decide the best way to display the stamps throughout the album. They use multiplication. This allows them to find out how many stamps are on their pages. Julia multiplies 9 × 4. Grandfather multiplies 6 × 6.

They both smile. Grandfather and Julia have different layouts, but the same number of stamps on each of their pages. They agree that 36 must be the perfect number of stamps for a page. They also agree to use Julia's layout for the album. Each page will have nine rows of four stamps each.

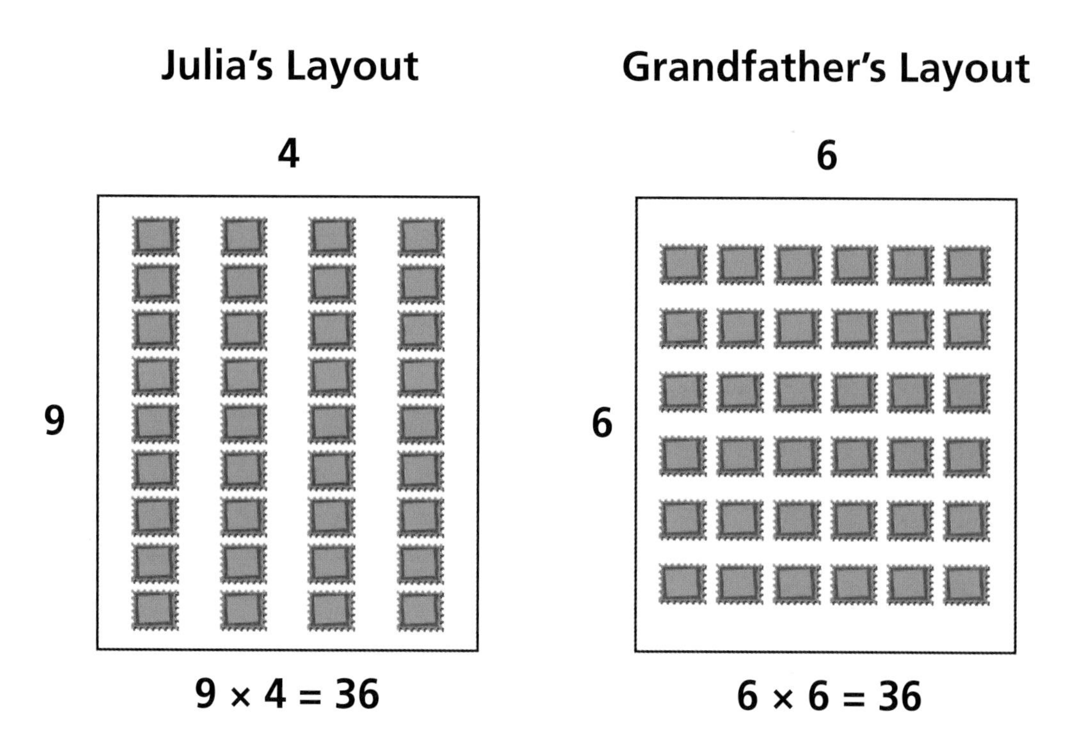

Chapter 2:

Stories from Stamps

Grandfather's collection has stamps from many countries. In fact, Grandfather thinks that 18 different countries are represented in his stamp collection. Some of his favorite stamps are from the United States. Grandfather finds a page that has many stamps from the 1996 Summer Olympic Games. These games were held in Atlanta, Georgia.

Grandfather is fond of these stamps. He tells Julia that he and Grandmother actually went to some of the Olympic events. Julia is surprised to learn that Grandfather has visited some of the places pictured on the stamps! Grandfather tells wonderful stories. Julia is eager to hear the stories about other stamps, too.

Grandfather continues. He shares his memories about the 1996 Summer Olympics while Julia looks at the stamps. Soon they get back to organizing the new album. Grandfather thinks there are enough stamps that they will fill two pages of the album for each country. Julia wonders if there are enough pages to hold all those stamps. Julia uses multiplication to find out.

$18 \times 2 = 36$

There are stamps from 18 countries. Each country will fill two pages in the album. They need 36 pages. The new stamp album has 40 pages. It will be perfect.

This photo album has enough pages for Grandfather's stamp collection.

This stamp commemorates the Mars Rover Sojourner mission.

Julia places stamps from the United States onto a page. She finds a stamp that says "Mars Rover Sojourner." Julia is curious. She learned about Mars in science class, but she doesn't know about the Mars Rover Sojourner. Grandfather says that the Sojourner was a robotic vehicle sent to Mars on the Pathfinder—an unmanned spacecraft. It was sent into space in 1996. Its mission was to collect data about the planet's rocks and soil. It also took thousands of pictures. The stamp was issued to commemorate the mission.

Julia enjoys her Grandfather's stories. She learns many interesting things from him.

Julia figures out how long the project will take.

Julia thought that organizing the stamps would be a fast project at first. She is quite surprised at how much time it is taking. She looks at the clock. Four hours have passed. Julia and Grandfather have completed two pages each. They have completed four pages altogether. Julia thinks for a moment. She figures out that it takes two hours to fill each page with stamps. With 36 pages to complete, the project will take about 72 hours. Julia does not mind at all.

Shortly, Julia and Grandfather decide it is time for a little break. They share a healthy snack. As they eat, they talk about the project.

Chapter 3:

Julia: The Newest Philatelist

Julia enjoys looking at Grandfather's stamp collection. She loves spending time with Grandfather. They are having fun.

Julia notices a special stamp in Grandfather's collection. She points to the stamp. The stamp is similar to a painting hanging in Grandmother's and Grandfather's house. Grandfather smiles. He explains that the picture on the stamp is a painting called "Big Raven." The painter, Emily Carr, was a well-known Canadian artist and writer. The stamp celebrates her life's work. Since Grandfather and Grandmother like Emily Carr's artwork, they bought one of her paintings years ago.

Julia wonders if some of the people, places, and events that she likes are pictured on stamps, too. She asks Grandfather that question. Grandfather thinks that there are many stamps with images Julia would recognize. Some might be in Grandfather's collection. There are new stamps issued each year, too. They might include people, places, and things Julia knows.

Grandfather suggests that Julia do some research. That sounds like a good idea to Julia. She decides that will be her next project. In fact, after she helpsGrandfather organize his stamps, Julia wants to start a stamp collection of her own.

Grandfather and Julia talk about stamps with familiar images.

Grandfather and Julia decide how to finish the album pages.

After their snack, Julia and Grandfather decide not to return to the stamp collection. Instead, they decide to split up the work that remains. They will work on the project during the week. Then they will meet again next weekend. At that time, they will see how much they have each accomplished.

Grandfather has a busy week planned. He will not have much time to work on the project. Julia is enjoying the project so much that she gladly agrees to finish a few extra pages. Before they clean up for the day, they decide how to split the remaining pages to finish the stamp album.

There are stamps from 18 countries. They completed album pages for two countries, Japan and Australia. They have some stamps from the United States in the album, too. There are stamps from 16 countries left to place in the album. Julia will complete album pages for ten countries. Grandfather will complete album pages for six countries.

Julia gives Grandfather the album pages he will need. Grandfather needs 12 album pages. That is enough for two pages for each of six countries. Julia needs 20 pages to finish her part of the project. That's enough for two pages for each of ten countries.

Julia's Album Pages

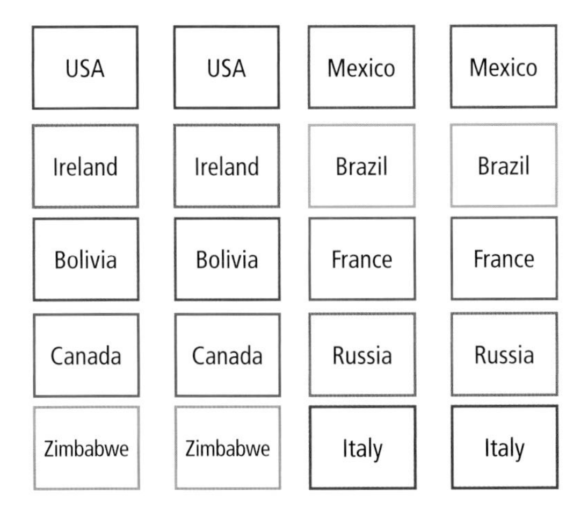

$10 \times 2 = 20$

Grandfather's Album Pages

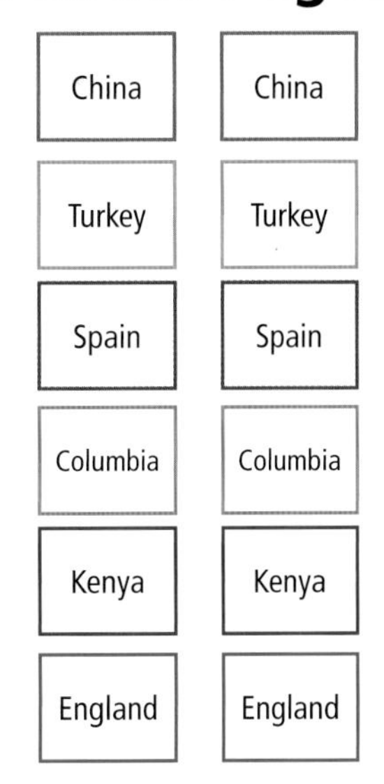

$6 \times 2 = 12$

Before he leaves, Grandfather reminds Julia to handle the stamps carefully.

Grandfather reminds Julia how to handle the stamps. "Remember, these stamps are not all expensive," Grandfather says. "But many of them have sentimental value." Julia understands. She knows that the stamps remind Grandfather of different people, times, and places. Julia promises to handle the stamps carefully.

Julia wonders. If she completes 20 album pages and each page holds 36 stamps, how many stamps will she place in the album? She multiplies 20 × 36. That's 720 stamps in all. Julia can't wait to look at all the stamps. When she sees Grandfather next weekend, she knows she will hear some more great stories.

Julia gives Grandfather a hug as they say good-bye. She thanks him for a special day. Just then, Grandfather reaches into his pocket. He hands Julia a postcard. On the postcard is a stamp from Greece.

"This is for you," Grandfather says. "There is a story behind this stamp, too. But, that will have to wait until the next time we get together. Until then, put this stamp in a safe place."

Julia smiles as she waves so long to Grandfather. She has the first stamp for her collection now. She can hardly wait to place it in her own stamp album.

This postcard from Greece includes the first stamp for Julia's collection.

Glossary

album a book with blank pages used to store things that are collected

arrange to place in order

commemorate to honor or remember

layout an arrangement or plan

philatelist a person who collects stamps

souvenir a small item that is a reminder of a place or an occasion